Understanding the Structure and Function of Proteins

– The Molecules of Life

Why I Wrote This Book

When Watson and Crick in 1953 discovered the double helix structure of DNA they claimed to have discovered the secret of life.

In biological terms there is much truth in that, because of course the structure of DNA tell us much about how it replicates and passes on information,

But let us not forget what the DNA code is actually coding for – proteins!

That's right – every single gene is a code that instructs the cell which of the 20 amino acids to put in which places in order to make the structure of that particular protein; be it insulin or amylase.

So – it could be argued that proteins are the molecules of life – the business end if you like – where the action really happens.

What I have done in this book is to explain in question and answer form, the structure and function of proteins in a clear and easily understood manner.

I know reading this book will improve your understanding of biochemistry and biology – and I am delighted to recommend it to you.

Book Contents

Chapter One

What Are Proteins?

How Important Are Proteins?

Proteins are arguably the most important of all biological molecules. They make up a large amount of the body's structure and biochemistry. For example, much of muscle and blood tissue is protein; all enzymes and antibodies are proteins, along with a number of hormones.

What Are Proteins Made Of?

Proteins contain mainly carbon, hydrogen, and oxygen and nitrogen atoms but they also often contain a few sulphur atoms as well.

Proteins are polymer molecules. They are made up of a long chain of different amino acids which have joined together by strong peptide bonds into what is called a polypeptide.

Also; many proteins have chemical groups attached to them called 'prosthetic groups' that are not made up of amino acids. These groups help the protein to perform its function.

What Are Amino Acids?

There are twenty different kinds of amino acid.
Every amino acid has the same basic structure, ending in a COOH group and NH2 group – so these can join together to form a strong and stable covalent peptide bond, linking the amino acids together.
The difference between amino acids is in the 'R' group which is different in each amino acid. The 'R' groups can have different chemical properties e.g. it might be a non-polar group such as CH3 or a polar group such as NH2.

What Are Globular Proteins?

The word 'globular' means 'ball-shaped' – so globular proteins are ball-shaped proteins.

The ball shape gives these kinds of proteins solubility - most of the globular proteins are soluble in water – and this is very important for their function as, for example; enzymes and blood proteins.
Some globular proteins need to dissolve in the cell membrane, and the globular shape helps with this too.

What Are Fibrous Proteins?

Fibrous proteins have a long and thin shape like fibers. They are strong, elastic molecules. This means they are able to form part of the body's structure, fibrous proteins include: collagen (which is found in tendons and cartilage); keratin (in hair and nails) and actin and myosin (in muscle).
These proteins tend to have a repeating amino acid structure and form many cross-links between their fibers, making them very strong. They resemble nylon in their structure.
They are not soluble in water.

How Big Are Protein Molecules?

Protein molecules can be very large and some are made up of hundreds of thousands of amino acids. However, on the other hand, some proteins are very small, such as insulin, which has only fifty-three amino acids.

How Many Different Proteins Are There In The Human Body?

In the human body there are more than twenty thousand different proteins. As each gene codes for a protein, this number is roughly the same as the number of genes in the human genome.

Chapter Two

The Structure of Proteins

What Is The Basic Structure of Proteins?

Proteins are biological polymers which are made up of at least one polypeptide chain that consists of a string of amino acids which are joined to each other by strong, covalent 'peptide' bonds.

Some proteins are made up of more than one polypeptide chain. An example is haemoglobin which is made up of four polypeptide chains. Insulin, while very small, is also made up of two separate tiny polypeptides which are joined together.

Some proteins also have non- amino-acid groups attached to them called 'prosthetic groups'. In the case of haemoglobin this is the ring structure and central iron atom comprising the 'haem' group.

What Are the Levels of Protein Structure?

There are four levels of protein structure.

The four levels are: primary, secondary, tertiary, and quaternary structure. A single protein molecule may contain one or more of these protein structure types.

What Is The Primary Structure?

The primary structure is simply the particular sequence of amino acids that makes up the polypeptide(s) in the particular protein. The first primary structure to be worked out was that of insulin, one of the smallest with 51 Amino acids: the first five of which are glycine, isoleucine, valine, glutamic acid, glutamine.

What Is The Secondary Structure?

The secondary structure takes one of two forms: either an alpha-helix or a beta-pleated sheet.

Both the alpha helix and beta-pleated sheet structures are held together by relatively strong hydrogen bonds between the C=O group of one amino acid and the N-H group of another.

Which structure is adopted depends on the amino acids present and their order in the primary structure.

What is the Tertiary Structure?

The tertiary structure is the exact three-dimensional shape which the secondary structure folds up into.

On a basic level it is either a roughly ball-shape for globular proteins such as enzymes and antibodies or a long fibre for fibrous proteins such as keratin and collagen.

However for each protein the tertiary structure is slightly different, allowing the specificity so vital to the function of proteins such as enzymes (where each enzyme has a different shape of active site; and therefore works with a specific substrate) and antibodies (where each antibody has a different shape of binding site and therefore binds to a specific antigen).

What is the Quaternary Structure?

Some proteins are made up of more than one polypeptide chain and these have a quaternary structure.

The quaternary structure is the result of one or more tertiary structures bind together, such as in haemoglobin, when four globular, tertiary structures join together to make one haemoglobin molecule.

What Are Prosthetic Groups?

Some proteins have prosthetic groups. These are chemical groups attached to proteins which are not made up of amino acids, in other words they are not part of the polypeptide chain but are attached to it.
Many proteins have prosthetic groups: e.g. glycoproteins - such as those found in the cell membrane - have monosaccharide sugars attached to them as prosthetic groups. Many enzymes have metal ions attached to them which help them to catalyse chemical reactions. Sometimes the

prosthetic group is a lipid, in which case it forms a lipoprotein. Many of these are also found in the cell membrane.

The 'haem' group is specific to the protein haemoglobin which is found in red blood cells. The haem group contains a ring with an iron atom in the middle of it – it is this iron atom that enables the haemoglobin to carry oxygen – by binding to it chemically at the lungs and releasing it at the respiring tissues.

Chapter Three

The Functions of Proteins

It is a good idea to approach proteins from a functional point of view.
Proteins have many different functions. But there is a common thread that runs through them. It is usually the case that proteins are used where there is a need for variability in structure. Thanks to the twenty different amino acids there is an almost infinite variety of primary structures, and hence secondary and tertiary structures (i.e. 3-d shapes) of proteins.
The function of proteins is very dependent on their structure.

How Do Proteins Function As Antibodies?

Antibodies are 'Y-shaped' proteins which are produced by cells called 'B plasma cells' (a type of white blood cell that forms part of our immune system) and which help to defend us against infecting pathogens such as bacteria and viruses.

They help to firstly, identify and then assist in destroying, these pathogens. Antibodies work in the blood, usually in conjunction with white blood cells, by binding to the antigens on the surface of a pathogen, immobilizing and labeling them for later destruction by white blood cells.

In order for this to happen antibodies must have a specifically shaped binding site that matches the shape of a specific chemical 'antigen' on the surface of the pathogen (usually on the outer surface of the cell wall or viral coat 'capsid'). This binding site is found at the top ends of the 'Y' shaped antibody molecule. It is known as the variable region of the molecule and is a different shape in different types of antibody that therefore can bind specifically to a different antigen.

The idea of specific immunity is that after an initial infection, the body is immune to a specific pathogen (bacteria, fungus or virus which causes a disease). Specific antibodies and the B cells which produce them (memory cells) remain in the blood and get rid of the specific pathogens that they recognize straight away, the second time they try and infect the body; meaning the person is immune to the particular disease.

How Do Proteins Function As Enzymes?

Enzymes are proteins that increase the rate of chemical reactions in the body. In fact, the necessary chemical reactions in the body would not happen fast enough without enzymes. It is no exaggeration to say that; without enzymes life on Earth would literally not be possible.

There are approximately twenty thousand human genes and around ten thousand; (so that is, half of the human genes); code for enzymes. The reason for this is that each enzyme is specific to only one chemical reaction, therefore each chemical reaction in the human body requires a specific type of enzyme to catalyse it. For example, in the digestive system, the disaccharide sucrose, can only be chemically split (digested) by the action of the enzyme sucrase (produced in the pancreas) and by no other enzyme.

The structure of enzymes allows them to be so specific for the particular reaction they catalyse because each type of enzyme (e.g. amylase) has a particular shape of active site which can only fit with and bind to the particular reactants (called substrates) involved in the specific chemical reaction that the enzyme catalyses.

How Do Proteins Act As Muscle Fibres?

The fibrous proteins actin and myosin combine to allow muscle cells to contract. Connections form between the fibres and a kind of rowing action moves the fibres together, shortening them – this is termed the 'sliding filament' hypothesis.

How Do Proteins Function As Hormones?

Some hormones are protein molecules. Probably the best known is insulin, which was actually the first protein to have its primary structure worked out, it only consists of 53?? amino acids, and as such is a very small protein indeed.

Insulin acts on the liver and muscle cells to instruct them to take up glucose from the blood and store it up as glycogen. Insulin is released from the beta-cells of the pancreas into the blood. It works by a typical negative feedback mechanism to maintain homeostasis of the level of glucose in the blood, so it does not rise or fall too much.

The particular shape of insulin means it will only bind to the cell membranes of certain target organ cells (such as those in the liver) – this is mediated by particular-shaped receptor molecules in the cell membrane surface which are often glycoproteins and lipoproteins.

Secretin, is another example of a protein hormone, it helps to control digestion by stimulating the pancreas and the intestine to produce digestive secretions.

Oxytocin, is another example. It stimulates the uterus to contract during birth.

How Do Proteins Function As Membrane Transport Proteins?

The cell membrane forms a barrier to most of the molecules that either flow in the blood plasma or are present in the cytoplasm. Within the cell membrane are globular proteins, floating in the 'sea' of phospholipids.

Proteins are well suited to this function as these transport proteins need to be a specific shape in order to transport a specific ions or molecule across the cell membrane: e.g. there is a specific carrier transport protein for glucose present in cell membranes.

Having specifically shaped transport proteins in the cell membrane means that the cell can select which molecules enter or leave the cell.

How Do Proteins Function As Blood Proteins?

There are a few different blood proteins which have different structures and functions:

- Albumin – this is a very large globular protein that helps to maintain the level of water in the blood. Its large size means that even though it is soluble in the plasma; it cannot move through the gaps in capillary walls and so remains in the plasma, helping to pull water back in and so preventing the accumulation of tissue fluid. When albumin levels are low, tissue fluid can accumulate around the belly and arms, giving the swollen belly appearance of kwashiorkor – a condition associated with protein malnutrition, where the body breaks down albumin to make up for a lack of amino acids in the diet.

- Haemoglobin – this is another large globular protein. It is found in red blood cells. It is made up of four separate globular polypeptide chains, each of which has a haem prosthetic group containing an iron atom at the centre. It is these iron atoms that allow haemoglobin to bind to oxygen at the lungs and carry the oxygen to the respiring tissues where it is released. The iron atoms each can bind to an oxygen molecule at the lungs and release it at the respiring tissues.

Haemoglobin is poisoned by carbon monoxide, which binds irreversibly (permanently) to the iron atoms in the haem group, preventing them from binding to oxygen.

- Fibrinogen – this is a globular protein that is involved in the complex blood clotting mechanism. Its role is to change into the insoluble fibrous form called fibrin, during the clotting sequence of reactions. Fibrin is an insoluble fibrous protein that catches platelets and red blood cells in its mesh, so helping to form the blood clot.

How Do Proteins Act As A Source Of Energy?

This is one function where the three-dimensional shape of the protein has no influence at all.

Protein is a major source of energy in most human diets. If you consume more protein than you need for body tissue maintenance and repair, your body (to be exact, the liver) will break it down into its constituent amino acids, remove the amine group from them (de-aminate them) and then either respire the remaining molecule to make energy or metabolize them into fat which is stored in the fat cells as a store of chemical energy.

The amine group is highly toxic and is made into urea (which is less toxic) in the liver and removed from the body in the urine.

Chapter Four

Proteins and Genetic Illnesses

Which Faulty Protein Causes The Genetic Disease, Cystic Fibrosis?

Cystic fibrosis is caused by a genetic defect which prevents the proper function of the CTFR protein which is a membrane transport protein that is present in the epithelial lining of the lungs, pancreas and testicles; helping to produce watery mucus.

When the CTFR membrane protein does not function properly, this leads to the production of thick, sticky mucus which: blocks up the bronchioles in the lungs – leading to difficulty breathing and frequent lung infections – and blocks the pancreatic duct – leading to poor digestion as few enzymes are released during digestion.

Unfortunately the poor function of lungs and pancreas can often lead to early death in cystic fibrosis sufferers, although with the help of modern therapy; many now live into their forties and fifties.

Which Faulty Protein Causes The Genetic Disease, Phenylketonuria?

Phenylketonuria is caused by a faulty gene that leads to the lack of function of an enzyme in the liver which (in healthy people) breaks down the amino acid phenylalanine.

Unfortunately, the non-function of the enzyme therefore leads to the build up of levels of phenylalanine in sufferers and this can lead to mental and physical deterioration.

The condition can be treated through control of the amount of phenylalanine consumed in the sufferer's diet. If the consumption of phenylalanine is kept down the level of phenylalanine should not rise too high in the body and the affected person should not suffer any ill affects of the disease.

Which Faulty Protein Causes The Genetic Disease, Albinism?

Albinism is caused by a malfunction of an enzyme involved in a one of a series of reactions which leads to the production of the chemical melanin – which gives colour to the skin and eyes and hair.

The lack of this enzyme's function therefore means that melanin is not produced.

Lack of melanin means that the eyes just show the colour of the blood vessels at the back of the eye (hence the eyes look pink) and the skin and hair are both colourless. This is the typical appearance of albinos. This genetic condition is very rare in humans, affecting only one person in millions.

Which Faulty Protein Causes The Genetic Disease, Sickle Cell Anaemia?

This is caused by a fault in the blood protein haemoglobin, found in red blood cells, which binds to oxygen at the lungs and transports it to the respiring tissues. The fault is due to the polypeptide chain having one amino acid different to the healthy form of the protein due to a single base; 'point mutation' in the gene which codes for this polypeptide.

This defect results in the 'R' group of this particular amino acid being non-polar and as it is on the outside of the haemoglobin molecule – it causes them to stick together rather than stay in solution - particularly at low oxygen concentrations. This causes the haemoglobin to form big chunks which are insoluble, and results in the red blood cells form a 'sickle' shape and become hard and inflexible - hence the name of the disease.

During exercise, when oxygen concentrations are low, the hard and inflexible sickle cells tend to block up the capillaries, causing pain due to a build up of pressure against the blockage (particularly in the muscles), breathlessness and some organ dysfunction.

Which Faulty Protein Causes The Genetic Disease, Haemophilia?

This is caused by a defect in the blood protein 'factor 8' which is involved in the very complex series of reactions that cause blood to clot when a wound in sustained.

The lack of function of factor 8 means that the blood will not clot. This is the typical symptom of haemophilia.

It is potentially a fatal condition because a serious wound can lead to the haemophilia sufferer bleeding to death, particularly in the past there were deaths.

It can now be treated with blood transfusions, or administering factor 8 to the sufferer. It need not be a fatal condition.

Because this gene is found on the X chromosome, what is called 'sex-linkage' - this condition only affects males.